Energy 131

树上长石油

Oil Growing on Trees

Gunter Pauli

[比]冈特·鲍利 著

[哥伦]凯瑟琳娜·巴赫 绘

贾龙智子 译

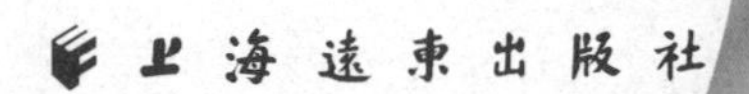

特别感谢以下热心人士对童书工作的支持：

匡志强　宋小华　解　东　厉　云　李　婧　庞英元
李　阳　梁婧婧　刘　丹　冯家宝　熊彩虹　罗淑怡
旷　婉　王靖雯　廖清州　王怡然　王　征　邵　杰
陈强林　陈　果　罗　佳　闫　艳　谢　露　张修博
陈梦竹　刘　灿　李　丹　郭　雯　戴　虹

目录

树上长石油 4

你知道吗? 22

想一想 26

自己动手! 27

学科知识 28

情感智慧 29

艺术 29

思维拓展 30

动手能力 30

故事灵感来自 31

Contents

Oil Growing on Trees 4

Did You Know? 22

Think About It 26

Do It Yourself! 27

Academic Knowledge 28

Emotional Intelligence 29

The Arts 29

Systems:
Making the Connections 30

Capacity to Implement 30

This Fable Is Inspired by 31

一棵巴西苦配巴树正在美国加利福尼亚州拜访一些亲戚。他们正享受太平洋海岸的壮丽景色。这是个适合放松的好地方，他们讨论起未来。

“谢谢你们欢迎我到你们家来。”苦配巴树说，“这是个美好的国家，但我很想知道，在进口这么多的水和燃料的情况下加利福尼亚州还能撑多久？”

A Brazilian tree, the Copaiba, is visiting some relatives in California. They are enjoying a magnificent view of the Pacific coast. It is a good place to relax and discuss the future.

“Thank you for welcoming me to your home,” the Copaiba says. “This is a wonderful country, but I do wonder how long California will survive, having to import so much water and fuel?”

……太平洋海岸的壮丽景色。

... a magnificent view of the Pacific coast.

……从拥有水和燃料的地方购买这些资源……

… buy water and fuel from those who have it …

“我们在这里挣了很多很多钱，美国可以称作世界上最富有的国家之一！我们可以从拥有水和燃料的地方购买这些资源，让人把它们运过来。”大戟科树回答道。

“好吧，你可能很富有，但是世界上会永远有足够的水和燃料来满足你们巨大并且日益增长的胃口吗？”

“We earn so much money here in the States, we could be considered one of the richest countries in the world! We can buy water and fuel from those who have it – and have people ship it to us,” responds the Eupho tree.

“Well, you may be rich, but will there always be enough water and fuel in the world to satisfy your big – and ever growing – appetite?”

“如果资源不够了，我们会种更多的灌木和树。”大戟科树回答，“然后我们会得到我们所需要的所有燃料。”

“为什么你们要等到为时已晚才行动呢？你我都知道，种下一棵树后，要花很多年才能收获。”

“If there is no longer enough, we will just plant more shrubs and trees,” responds the Eupho, “and we will have all the fuel we need.”

“Why would you wait until it is too late? You and I both know that when a tree is planted, it takes years before you can harvest it.”

……我们会种更多的灌木和树……

… we will just plant more shrubs and trees …

……新鲜柴油……

... fresh diesel fuel ...

“你知道，你完全是正确的。人类消耗了太多燃料，长途输送了太多的水。而事实也确实如此：他们的确倾向于到了最后一刻才行动。”

“生活在亚马孙河流域的人已经通过钻孔利用我们很多年了。”苦配巴树说道，“他们在我的树干上钻一个小孔，从那里流出的新鲜柴油可以直接加入引擎里使用。”

“You are right on all counts, you know. People consume too much fuel and pump too much water over long distances. And it is true: they do tend to act only when it is too late.”

“The people living in the Amazon have been tapping me for ages,” Copaiba says, “They drill a little hole in my stem and the fresh diesel fuel that flows out can be put straight into an engine.”

“你是说他们停在树旁加油而不是去加油站？这只是个空想——但是挺有意思！好莱坞可能会立即拍一部这样的电影。”大戟科树笑道。

“你可以尽管笑，但是人类最好先设计出燃油利用效率极高的汽车，而不是他们在这里开的这种油耗量大的车。我们每棵树每次钻孔只能提供二十升油，一年里也许最多只能钻两次。”

“You mean instead of going to the gas station, they stop by a tree and fill up? That is just a fantasy – but an interesting one! Hollywood would immediately make a movie about that,” laughs Eupho.

“You can laugh all you like, but they had better first design cars that are very fuel efficient, instead of these guzzlers they drive around here. We can only offer twenty litres per tap, and then only maybe twice a year.”

……停在树旁加油？

... stop by a tree and fill up?

……我比你含有更丰富的燃料。

... I am richer in fuel than you are.

“这样啊，我比你含有更丰富的燃料。”大戟科树自夸道，“而且我不需要在郁郁葱葱的亚马孙雨林里生长。我能在干旱的土地上扎根——就在加利福尼亚这里。”

“只要能利用当地的植物——特别是像你这样友好的——发展出具有当地特色的解决方案，那就尽一切可能去做吧！没必要向海外寻求机会。”

“Well, I am richer in fuel than you are,” Eupho boasts, “And I do not need to be in any lush rainforest in the Amazon to grow. I am able to grow on dry land – right here in California.”

“Whenever one can have a local solution using a local plant, especially a friendly one like you, by all means, go for it! There is no need to look overseas.”

“你知道吗？我们大戟科树每公顷可以生产十五桶石油和十五桶乙醇，剩下的生物质足以生产更多能量。”

“如果你们真的有能力提供燃料，而且你们能生长在不毛之地，为什么你们的燃料没有被大家使用呢？”

“Did you know that we euphos can produce fifteen barrels of oil and fifteen barrels of alcohol per hectare, and the biomass left can generate all the power needed to produce more energy.”

“If you really are this good at providing fuel, and you can grow on land where nothing else will grow, why is your fuel then not being used by everyone?”

……剩下的生物质足以生产更多能量。

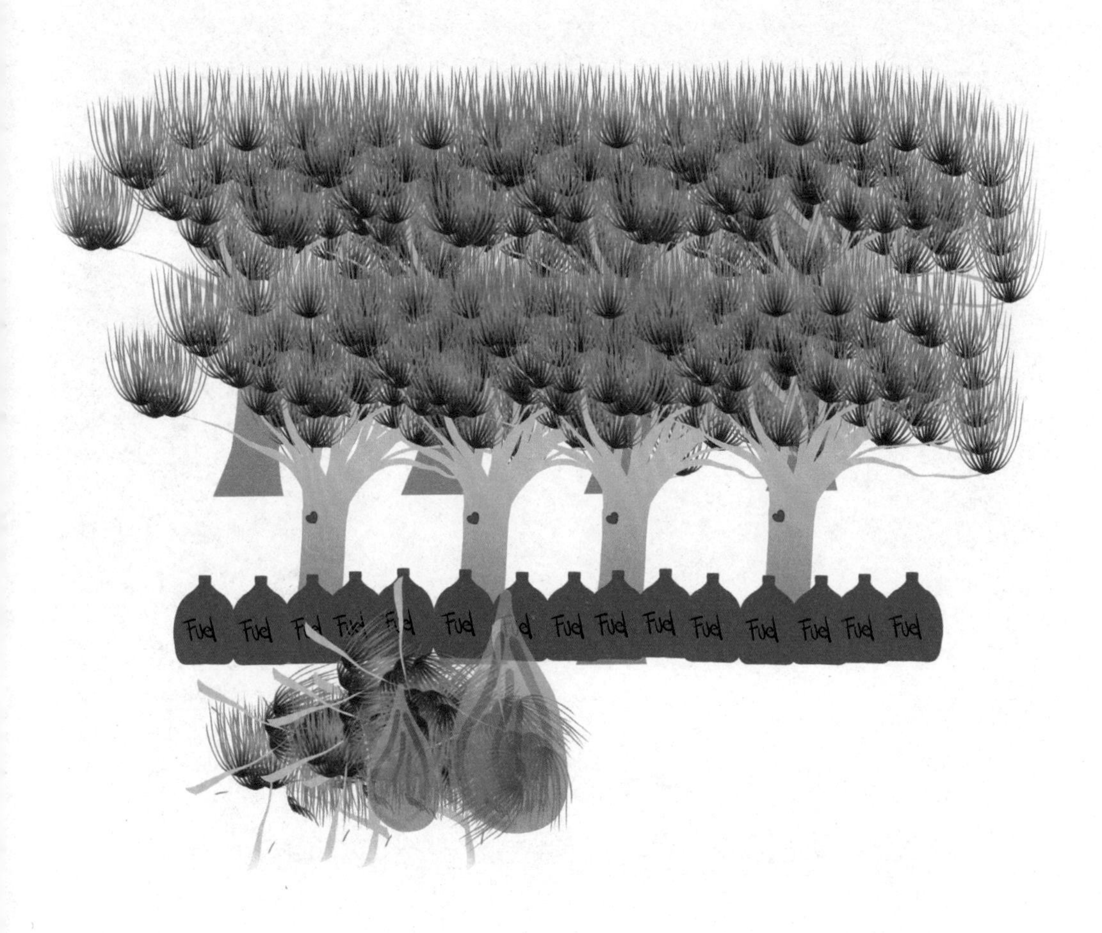

... biomass left over can generate more energy.

……海平面上升和气候变化吗？

... rising sea levels and climate change?

“我经常听到别人问这个问题！你认为那些现在通过燃料赚了大钱的人们是怎么想这个问题的呢？”

“可以肯定的是，那些通过贩卖石油发财的有钱人无法阻止你们提供燃料，因为你们对大自然物尽其用，又不会破坏地球。他们真的不关心海平面上升和气候变化吗？”苦配巴树很想知道。

“I have heard this question asked so often! What do you imagine those people, who make a lot of money from fuel now, think about this?”

“But surely all those rich people, who make a fortune selling petroleum, cannot stop a fuel like yours, one that uses the best Nature has to offer, without harming the Earth. Do they really not care about rising sea levels and climate change?” Copaiba wants to know.

“嗯，看起来他们确实不关心。他们可不愿意用得克萨斯的油井来交换加利福尼亚或巴西的植物。”

“可能他们这一辈人不关心，但是我确定他们的孩子或孙辈会关心的。”

……这仅仅是开始！……

“Well, it seems that they do not. They are just not willing to trade their Texan oil wells for Californian or Brazilian plants.”

“It may be that the older generation does not care, but I am sure their children and grandchildren do.”

… AND IT HAS ONLY JUST BEGUN! …

……这仅仅是开始！……

... AND IT HAS ONLY JUST BEGUN! ...

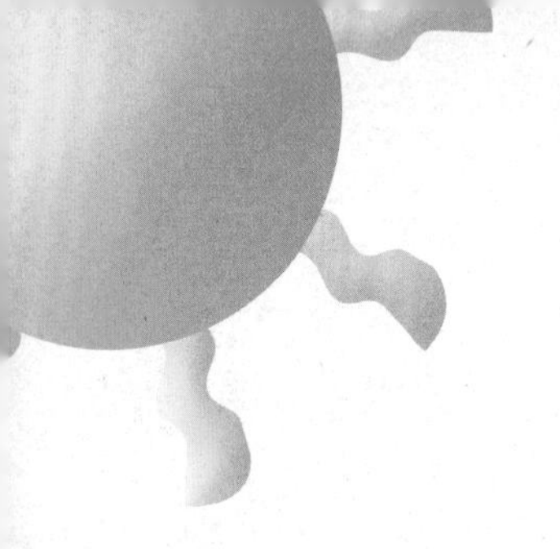

你知道吗？

300

= 12 000 L

A mature fuel tree yields up to 40 litres of oil per year as a result of a fungus. One hectare with 300 trees provides 12,000 litres of free fuel, good for driving 200,000 km, with a fuel efficiency of 6 litres/100 km.

一棵成熟的燃料树在真菌作用下每年能产出40升左右的石油。一公顷土地的300棵树能提供12 000升免费的燃料，足够燃料效率为6升/百千米的车开20万千米。

The Copaiba tree has a lifetime of 70 years. Its flowers attract stingless honeybees, which provide an excellent medicine.

苦配巴树能活70年左右。它的花吸引无针蜜蜂，并且具有药用价值。

Pine tree's sap is used to make turpentine which is commonly used to clean paintbrushes. It was used to power motorbike engines in Japan after the Second World War, but too many impurities in it caused too much smoke.

松树的树液可制成松节油，普遍用于清洁画笔。第二次世界大战之后，日本曾经用它给摩托车发动机提供动力，但是它含有的过多杂质会引起太多烟尘。

Pine trees also produce a fuel, but it is not pure and ready to use. It is a by-product from the resin harvested for the paint industry.

松树也能产生燃料，但是并不纯净，也不能直接使用。它是采集用于涂料行业的树脂时的副产品。

Turpentine is a zero emission fuel, meaning that even when you burn it, the carbon released by incineration through a combustion engine, is less than the carbon sequestered in the trees.

松节油是一种零排放燃料，这意味着从内燃机中燃烧燃料释放的碳比树木里固定的碳要少。

The leaves of the fuel tree cause skin irritation. Its latex is poisonous to humans and most livestock. Goats are however immune to the toxins and able to eat it.

这种燃料树的叶子会刺激皮肤。其乳胶对人类和大多数家畜来说是有毒的。然而山羊对这些毒素免疫，并且能够食用这些叶子。

The Californian fuel tree is known as a mole plant. Moles avoid burrowing around this tree, as they do not like the scent emitted by the roots, which affects the taste of the worms they prey on.

加利福尼亚州的燃料树被认为是鼹鼠的克星。鼹鼠会避免在这种树周围打洞，因为它们不喜欢树根散发的气味，这种气味还会影响鼹鼠抓到的虫子的口感。

Many innovations never enter the market or become mainstream. Those who benefit from the old and lesser efficient processes and technologies do not allow innovators access to the market.

很多创新的事物从没有进入过市场或成为主流。那些从守旧的、更低效的加工过程和技术中获利的人会阻挠创新者进入市场。

Do you think we could ever have enough fuel from trees to power all the engines we have?

你认为我们有可能从树木里获得足够发动我们所有引擎的燃料吗？

Will the oil barons ever be ready and willing to substitute fossil fuel with the fuel from the trees?

石油大亨们会准备好并且愿意用树木里的燃料取代化石燃料吗？

Is it fine that those who are rich consume as much as they want? Or, should everyone monitor their intake and reduce it, so more people can enjoy what the Earth produces, sustainably?

那些富裕的人尽情消费是对的吗？或者说，是否每个人都应该监测自己的摄取量并做出缩减，从而让更多人能可持续地享受地球的资源？

Should we take future generations into account when we make decisions now?

我们在当下做出决策的时候，是否应该考虑子孙后代？

Do It Yourself!

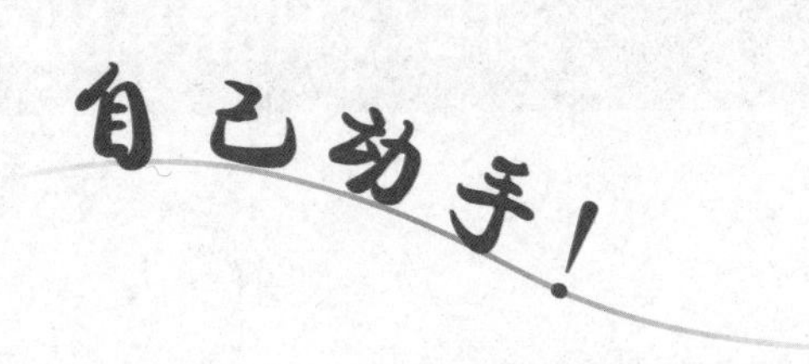

Have a good look at biodiversity around where you live. Is there any plant or tree that could produce latex or a resin that contains turpentine? How far do you have to travel to get to such trees? Draw a circle around your village or city and calculate the radius of how far one would need to travel to find any available fuel trees that could power engines without causing any emissions. If there are no such trees around where you live, what would it take to plant some?

Now construct a logical argument why such trees should be planted and point out the side benefits, such as ... no more moles!

好好观察一下你住所周围的生物多样性。有没有能生产含有松节油的乳胶或树脂的植物或树木？你要走多远才能找到这种树？以你所在的村庄或城市为圆心画圆，计算要找到燃料树的半径距离（它们能提供零排放的燃料）。如果你的住所周围没有这样的树，怎样才能种一些？

现在构建一个有逻辑的论点，证明为什么应该种这样的树，并指出附加的收益，比如……再也不会有鼹鼠！

学科知识

Academic Knowledge

生物学	生物分类学：科、亚科、族、亚族、属、物种；大戟科植物和苦配巴树之间的相似之处和不同点；对一种生物来说有毒的东西，对另一种来说可能是营养素；真菌（粉红粘帚霉）在柴油生产过程中对树木（心叶船形果木）里的纤维素进行加工；藻类能够生产石油和燃料。
化 学	碳氢化合物植物的化学组成；萜类化合物；萜烯分解为甲醇；植物可合成类异戊二烯；树木提供木材（砍树）、化学物质（采用树内汁液）和糖分（通过发酵）作为燃料；来自树木的不同燃料：芳族化合物、液化石油气、低相对分子质量燃料气和焦炭；微生物柴油的制造；石油和燃料的不同之处。
物 理	树脂通过蒸汽分解为松香和松节油；过滤松节油和乳胶将杂质析出，或者换一种方式，将液体倒出并在重力作用下将杂质留在底部，这些杂质可以通过别的方式移除。
工程学	设计燃油效率高的汽车；松节油可以带动汽油和柴油引擎。
经济学	石油市场的价格波动；资本密集行业；价格需求；碳排放总量和控制交易；燃油效率对竞争的影响；使用价格不高的土地并将其转化为无价的资源。
伦理学	通过出售石油发财，不处理相关的环境风险；在不损害大自然的条件下利用大自然最好的财富；考虑今天所做的决策和这些决策对未来7代人的影响；规划的重要性：可使群落的适应性更强；用粮食作物生产燃料与用纤维素废弃物生产燃料的机会；大量种植外来物种对本地生态系统造成的破坏。
历 史	亚马孙河流域原住民给树开孔获取燃料的传统；19世纪中叶在巴西亚马孙地区的橡胶热；对给树开孔来生产燃料的最早记录是在1635年的英格兰。
地 理	苦配巴树原产于巴西、阿根廷、委内瑞拉和巴拉圭；加利福尼亚是美国的一个州，但是就对世界经济的影响而言，它是最重要的地区之一；亚马孙地区；巴塔哥尼亚北部树菌根的生长。
数 学	对直径、长度和体积的计算；基于每公顷的升数计算产量。
生活方式	民间医学：使用植物来治疗创伤、炎症、皮肤溃疡、牛皮癣、蚊虫叮咬，减轻疼痛感；在化妆品行业使用油性树脂来制备香水、肥皂和化妆水；为“穷困的日子”储备金钱和资源。
社会学	好莱坞对社会的影响；需要找到基于时间和地点的解决方式来影响当地的生物多样性；亚马孙原住民和他们与森林之间的相互作用。
心理学	为什么人类要到了最后一刻才行动？预防是一种风险管理的方法。
系统论	在别的东西都无法生长的地方种植每公顷能生产15桶石油和15桶乙醇的树木，剩下的生物质可以用于产生生产燃料所需的电力。

情感智慧
Emotional Intelligence

苦配巴树

苦配巴树对受到的欢迎表示感激，并称赞这个地区的美丽。然而，苦配巴树很快发现这个地区消耗了大量的水和燃料。苦配巴树想知道，从需求不断增长的观点来看，是否有一天水和燃料都会不再充足了。他也担心只有当危机深重的时候人们才会得到教训。苦配巴树愿意分享从给树木开口取液到燃料效率的各种解决方案，敦促人们寻找本地的解决方案。苦配巴树鼓励加利福尼亚人也采用这种做法，但是对这方面的进展太小感到很惊讶。苦配巴树询问一场危机必然袭击整个社会的原因，建议下一代人也参与到决策过程中来。

大戟科树

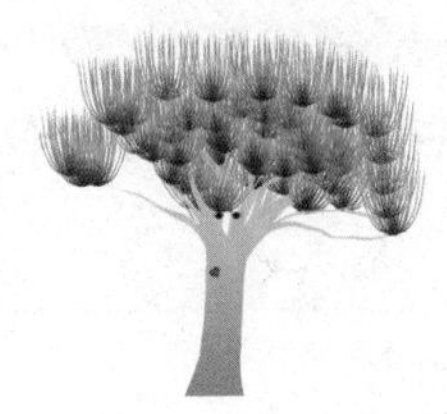

加利福尼亚州大戟科树夸耀金钱的力量以及当地居民能够买到他们想要的任何东西的能力。大戟科树很有信心地认为，如果资源不够，社会会采取纠正措施。但是，面对苦配巴树，大戟科树接受消费过高这一观点，并承诺在为时已晚之前采取行动纠正这一形势。大戟科树认为苦配巴树提出的解决方案是一种幻想，但是很欣赏这个故事，并认为这会成为一部优秀的好莱坞电影。这一对话引发了对一种已知解决方案的再探索，有现成的知识和充裕的土地来实施。但是大戟科树表现得很现实，指出阻力来自那些现在控制市场的人。

艺术
The Arts

研究这两种树的形状。拿一支铅笔，画出这两种树的特征。观察每棵树的轮廓，从树根开始看起，注意两种树的区别。然后仔细看看叶子。这些叶子有什么不同？继续只画出叶子的外轮廓。你现在拥有的是两种最多产的燃料树的图像——只靠一张纸、一支铅笔和你的一些近距离观察！

思维拓展

Systems: Making the Connections

我们需要满足整个世界对燃料的需求。农业是能源消费大户之一，我们都知道没有了水，我们将无法维持农业所需的生产效率水平。水和燃料决定了一个社会的恢复力。当需求增加、成本上涨的时候，那些买得起这些的人将很容易适应变化，而不能适应的那些人会蒙受损失。供水能力的下降与森林覆盖的减少有关。在我们已经将地球的植被覆盖减少到其原有的一半的时候，大家已达成共识，即应该种树，并且应该对失去了表层土的土地进行恢复性覆盖以提高土地保水能力。此外，如果我们打算审视当地的生物多样性，并寻找燃料提供水平高的树木时，应该将必需的森林覆盖率和能源生产结合起来。在树林的燃料生产力水平被认为很低，而且树木长成时间很长的情况下，那些以过度消费为特征的社会应该开始为后代考虑。当少数人认识到需要运用已有的资源（比如荒地）行动时，现有的权力者可能永远不会让更佳的替代方式成为主流，因为这将动摇他们已有的盈利模式。这种变革的阻力可能只符合当代人的利益。因此通过协商而制定决策是很重要的，与那些代表后代人的人们，或与那些有能力的人们进行协商。在修复森林时，如果我们选择本地的、能够在荒地上生长并产生燃料的树来进行重新种植，那么我们就有了系统的解决方案。

动手能力

Capacity to Implement

计算一棵树可以产生多少能源。首先，把树砍倒就能获取木材。也可以对灌木进行修剪，促进树的生长，在产生生物质的同时收获葡萄藤、茶叶或咖啡。木材是一种能够燃烧从而提供热量的燃料。木材也可被转化为能产生更多热量的木炭。还可以在树上开口获取乳胶和树脂，再将其转化为几种化合物。修剪后最终没变成木炭的废弃物以及树木开口取出的天然成分加工后的残留物，可在沼气池中发酵来生产沼气，在已经数量可观的燃料基础上增加了更多燃料。树木还能帮助保持土壤中的水，这样可以减少必需的抽水。你能算出产生和节约的能源有多少吗？这一切现在必须转化成实际的行动计划。我们有不止一种产生动力的方式——你来计算具体数目，得出能让政策制定者做出负责任的决策的论据。

故事灵感来自

This Fable Is Inspired by

梅尔文·卡尔文

Melvin Calvin

梅尔文·卡尔文生于美国明尼苏达州的圣保罗，其父母为俄罗斯移民。他拥有化学专业的博士学位，并曾在加利福尼亚大学伯克利分校担任教授。他创立了生物有机化学专业。他是光合作用碳减排方面的先驱。他的工作涉及各种学科，比如物理、生物和化学。这位 1961 年的诺贝尔化学奖得主解释了植物如何利用来自太阳的能量产生能源和食物。结合 20 世纪 60 年代中期不同背景的科学家的聪明才智，他对不同种类的树和灌木生产石油、乙醇和生物质的能力进行了系统研究。

图书在版编目(CIP)数据

冈特生态童书.第四辑：修订版：全36册：汉英对照 / (比)冈特·鲍利著；(哥伦)凯瑟琳娜·巴赫绘；何家振等译.—上海：上海远东出版社，2023
书名原文：Gunter's Fables
ISBN 978-7-5476-1931-5

Ⅰ.①冈… Ⅱ.①冈… ②凯… ③何… Ⅲ.①生态环境-环境保护-儿童读物—汉、英 Ⅳ.①X171.1-49

中国国家版本馆CIP数据核字(2023)第120983号

著作权合同登记号图字09-2023-0612号

策　　划 张　蓉
责任编辑 曹　茜
封面设计 魏　来李　廉

冈特生态童书
树上长石油
[比]冈特·鲍利　著
[哥伦]凯瑟琳娜·巴赫　绘
贾龙智子　译